I0504382

THE FAST LAPLACE
TRANSFORM

This monograph reviews the use of the Laplace transform as implemented numerically using the fast Fourier transform. This method has been described earlier by investigators in the electrical power community, but it does not seem to be widely used in the electromagnetic compatibility area. The goal in developing this monograph is to bring this computational method to the attention of the workers in this community by providing several examples and comments on its use for practical problems.

Frederick M. Tesche received his BS and Ph. D degrees in Electrical Engineering in 1965 and 1971, respectively, from the University of California at Berkeley. He is a consultant in the fields of numerical electromagnetics and electromagnetic compatibility (EMC). He is also an Adjunct Professor of Electrical Engineering at Clemson University.

Pierre F. Bertholet is an electrical engineer with an EE degree in 1976 from the Technical University of Yverdon, in Switzerland. He presently is working as a specialist in the EMC field in the Procurement, Technology and Real Estate Center of *armasuisse*, which is part of the Swiss the Federal Department of Defence, Civil Protection and Sports.

THE FAST LAPLACE TRANSFORM

Frederick M. Tesche

Holcombe Dept. of
Elec. and Comp. Engin.
337 Fluor Daniel Building
Clemson, SC 29634-0915

Pierre F. Bertholet

armasuisse HPE Laboratory
3700 Spiez, Switzerland

THE FAST LAPLACE TRANSFORM

First Edition: 2010

Copyright © 2010 by Frederick M. Tesche

All Rights Reserved.

This monograph, or parts thereof, may not be reproduced in any form without permission of the authors.

Printed in the United States of America by Lulu Enterprises, Inc.

http://www.lulu.com

ISBN 978-0-557-84705-1

This book is dedicated to the memory of Dr. Carl E. Baum
(1940 – 2010).

Preface

About 6 years ago, one of the authors (PFB) was faced with the requirement of performing spectral calculations for ill-behaved waveforms that cannot be easily treated using the Fast Fourier transform. Examples of this type of waveform include a unit step function and a semi-infinite sine wave.

He had heard anecdotal stories of some researchers modifying this type of waveform by multiplying it by an attenuating exponential function $\exp(-\sigma_o t)$ to artificially dampen the waveform so that a numerical Fourier transform could be taken. Then, after manipulating the computed spectral response (say, by applying a known transfer function, or by time shifting the spectrum), an inverse Fourier transform is performed. The resulting transient waveform is then multiplied by $\exp(+\sigma_o t)$ to remove the effects of the artificially induced damping function.

In thinking about this procedure, he realized that this is nothing more than a description of the *Laplace transform* of the function $f(t)$, which yields the spectrum not at a frequency $f = \omega/2\pi$, but at a *complex* frequency $s = \sigma_0 + j\omega$. Bertholet went on to develop a simple direct and inverse Laplace transform routine that uses the FFT algorithm to perform the necessary calculations. This concept was documented internally for his laboratory in French, but it was never published in the open literature.

Two papers in the electrical power community in 2004 and 2007 have described this computational procedure and have called it the *numerical Laplace transform* (NLT). These references have provided a historical review of the past work leading to the realization of this approach, and have given several illustrations of the use of this method for power system analysis. This method is actually based on an earlier paper in 1968, which refers to the method as the *modified Fourier transform* (MFT).

These earlier references provide a review of the theoretical basis for the method of computing the Laplace spectrum of $f(t)$ and in computing its inverse. However, the details of how the Laplace spectrum is to be used for practical problems are not discussed explicitly in this earlier work. This monograph serves to fill this void.

This monograph reviews the use of the Laplace transform as implemented using the fast Fourier transform. While this method is used in the electrical power community, it is not widely used in the electromagnetic compatibility (EMC) area. The goal in developing this monograph, therefore, is to bring this computational method to the attention of the workers in the EMC community by providing several examples and comments on its use for practical problems.

We are indebted to the Swiss Federal Office for Civil Protection of the Federal Department of Defense, Civil Protection and Sports for their support of the investigations leading to this document. In particular, we would like to acknowledge discussions and comments from Dr. Alain Jaquier and Mr. Markus Nyffeler, both from the Swiss **armasuisse** organization.

P. F. Bertholet, Bern, Switzerland
F. M. Tesche, Saluda, NC, USA

November, 2010

Contents

Figures

Chapter 1. Introduction

The numerical Fourier transform and its inverse transform are powerful analysis tools often used in the electromagnetics (EM) area. The development of the fast Fourier transform (FFT) by Gauss circa 1805 [1] and the subsequent machine implementation by Cooley and Tukey some 165 years later [2], have contributed greatly to our ability to pass from the time domain to the frequency domain, and back, with very large data records.

There are limitations, however, to Fourier analysis methods, which are usually encountered by anyone who tries to use them. A sufficient condition for a Fourier transform of a time-domain function $f(t)$ to exist, is that $f(t)$ must be absolutely integrable [3]:

$$\int_{-\infty}^{\infty} |f(t)| \, dt < \infty . \tag{1}$$

This condition is satisfied by all practical transient signals that have a finite duration[1]. However, some idealized waveforms, like the step function and the semi-infinite sinusoidal waveform, pose difficulties for Fourier transformation.

In engineering applications, the Fourier transform of a waveform can be computed either by a direct evaluation of the Fourier integral, or by the use of the discrete Fourier transform (DFT). In this latter approach, the transient waveform and the resulting spectral response are sampled at discrete time and frequency points. The FFT is a special case of the DFT, in which the number of points in the transient waveform, N, is constrained to be equal to $N = 2^m$, where m is an integer.

---

[1] As noted in [3], (1) is a sufficient, but not necessary condition for the existence of the Fourier transform of $f(t)$. Papoulis give as an example the function $f(t) = \sin(\omega_o t)/t$, which does not satisfy (1), but which does have a Fourier transform.

In practical cases where numerical evaluations are used, the time window of the transient signal must be truncated. In addition, the discrete sampling of $f(t)$ implies that the Fourier spectrum is band limited. Thus, the numerical transform is only an approximation to the actual Fourier transform of a waveform. If the sampling of f(t) and the time window in which the waveform is defined are not properly chosen, the computed spectrum can be incorrect.

These observations have led to two practical guidelines for performing numerical Fourier transforms, which are

- Choose the time window in which the waveform $f(t)$ is defined so that the waveform is zero at each end of the window, and

- Choose the number of points in the discrete sampling of $f(t)$ so that the rate of change of the fastest part of the waveform is adequately represented.

While these are seemingly rather straightforward requirements, they occasionally pose problems for computing the transform of certain types of functions. Several of these functions are shown in Figure 1. This figure shows a commonly used model for the return stroke current in a lightning channel and the resulting E-field [4], a step function and a semi-infinite sinusoidal signal. Each waveform is started at a different time for clarity, and the amplitudes are in arbitrary units.

In addition to these signals, waveforms that are distributions, such as the Dirac delta function $\delta(t)$ or $f(t) = 1/\sqrt{t}$, pose problems with the FFT because it is impossible to represent them numerically as discretely sampled functions[2].

---

[2] Of course, these functions also can be handled analytically.

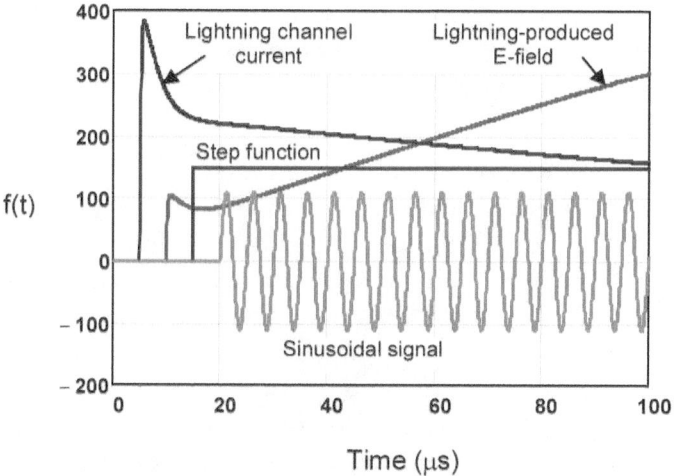

Figure 1. Illustration of several simple waveforms that pose difficulties in computing the Fourier transform using the FFT.

To perform spectral calculations of ill-behaved waveforms like those in Figure 1, it is possible to use the same method that is employed with analytically passing from the Fourier transform to the Laplace transform. That is, we can modify the initial waveform by an attenuating exponential function $\exp(-\sigma_o t)$ to artificially dampen the waveform in the time window so that a Fourier transform can be taken. Then, upon the manipulation of the computed spectral response, an inverse Fourier transform can be performed. The resulting transient waveform is then multiplied by $\exp(+\sigma_o t)$ to remove the effects of the initial artificial damping.

This procedure is essentially a description of the Laplace transform of the function $f(t)$, which yields the spectrum, not at a real frequency $f = \omega/2\pi$, but at a *complex* frequency $s = \sigma_o + j\omega$. With this method, it is possible to develop a simple direct and inverse Laplace transform routine that uses the FFT algorithm to perform the necessary calculations.

Recently, two papers in the electric power community have described this computational procedure, and have referred to it as the *numerical Laplace transform* (NLT) [5, 6]. These references have provided a historical review of the past work leading to the realization of this approach and have given several illustrations of the use of this method for power system analysis. This method has its roots in an earlier paper [7], which calls this method the *modified Fourier transform* (MFT).

These earlier references provide a good review of the theoretical basis for the method of computing the Laplace spectrum of $f(t)$ and in computing its inverse. However, the details of how the computed spectrum should be used for practical problems are not discussed explicitly. For example, if a signal $f(t)$ is passed through a linear system (say a filter) its Laplace spectrum must be multiplied by the Laplace spectrum of the system's transfer function evaluated at the complex frequency (s) rather than at the real frequency (f) in the Fourier domain. While this does not pose a problem for transfer functions that are known analytically, care must be used if the transfer function is *measured* in the Fourier domain. Such transfer functions in the Fourier domain must be analytically continued into the complex frequency domain to be used with the Laplace transform.

While the Laplace transform method of [5, 6] appears to be well-known to the power community, a poll of investigators in the EMC area suggests that this method is not commonly used by them, and that many people are unaware of its benefits.

In this monograph, we will refer to this numerical procedure as the *fast Laplace transform* (FLT) to be consistent with the notion of the Laplace transform as commonly used in the electrical engineering literature. We describe this method as "fast" because the FFT algorithm is used in the process.

Chapter 2. Theoretical Development

2.1. The Fourier and Laplace Transforms

For a transient waveform $f(t)$ that is zero for $t < 0$, its Fourier transform $F(\omega)$ is given by [3] as

$$F(\omega) = \int_0^\infty f(t)e^{-j\omega t}dt \ . \tag{2}$$

As noted earlier, a sufficient (but not necessary) requirement for the existence of $F(\omega)$ is given by (1). The inverse Fourier transform, which provides the transient waveform from the spectrum, is given by

$$f(t) = \frac{1}{2\pi}\int_{-\infty}^{\infty} F(\omega)e^{j\omega t}d\omega \tag{3}$$

For waveforms that do not meet the requirement of (1), yet which are exponentially bounded, a spectrum can be computed by multiplying $f(t)$ by an exponential damping function to create a *modified* waveform $f_m(t;\sigma_o)$ as

$$f_m(t;\sigma_o) = f(t)e^{-\sigma_o t} \ . \tag{4}$$

In this manner, the spectrum of this modified waveform becomes

$$F_m(\omega;\sigma_o) = \int_0^\infty \left(f(t)e^{-\sigma_o t} \right)e^{-j\omega t}dt$$

$$= \int_0^\infty f_m(t;\sigma_o)e^{-j\omega t}dt \ , \tag{5}$$

where f_m is a function of both time and the damping constant σ_0. In this expression, we note that the modified waveform f_m and the spectrum F_m are related by the Fourier transform of (2).

The inverse Fourier transform of the spectrum F_m provides the modified waveform f_m as

$$f_m(t;\sigma_o) = \frac{1}{2\pi}\int_{-\infty}^{\infty} F_m(\omega;\sigma_o)e^{j\omega t}d\omega \qquad (6)$$

and the transient function $f(t)$ may be reconstructed by multiplying f_m by $e^{\sigma_o t}$:

$$f(t) = e^{\sigma_o t}\frac{1}{2\pi}\int_{-\infty}^{\infty} F_m(\omega;\sigma_o)e^{j\omega t}d\omega \ . \qquad (7)$$

With $s = \sigma + j\omega$ and $F_m(\omega;\sigma_o) \equiv F(s)$, (5) and (7) form the well known Laplace transform pair[3]

$$F(s) = \int_{0}^{\infty} f(t)e^{-st}dt \ , \text{ and} \qquad (8)$$

$$f(t) = \frac{1}{2\pi j}\int_{\sigma_o - j\infty}^{\sigma_o + j\infty} F(s)e^{st}ds \ . \qquad (9)$$

Note that (8) is the so-called one-sided, or unilateral, Laplace transform, and arises due to the fact that $f(t)$ for $t < 0$ is assumed to be zero. A more general expression for this transform exists in the form of the two-sided, or bilateral Laplace transform, as discussed in [8]. In this monograph, however, we will use only the one-sided

---

[3] The reader is cautioned that there is a possibility of confusion here with regard to the notation for the Fourier and Laplace transforms of the function $f(t)$. In this monograph, and in the engineering literature, the function $F(\omega)$ or $F(f)$ (where $\omega = 2\pi f$) is commonly used to represent the Fourier transform of (2), and $F(s)$ (where $s = \sigma_0 + j\omega$) denotes the Laplace transform of Eq(8).

transform, as all practical signals of interest are zero prior to some "turn-on" time, which we take to be at $t = 0$.

The inverse Fourier transform of $f(t)$ in (3) and the inverse Laplace transform of (9) involve integrals along the Bromwich contours shown in Figure 2 for $\sigma = 0$ and $\sigma = \sigma_o$, respectively. While these integrals are formally defined for infinite limits of integration $(j\omega \rightarrow \pm\infty)$, these integrals are usually approximated by integrals with limits $(j\omega \rightarrow \pm\omega_m)$ as noted in Figure 2.

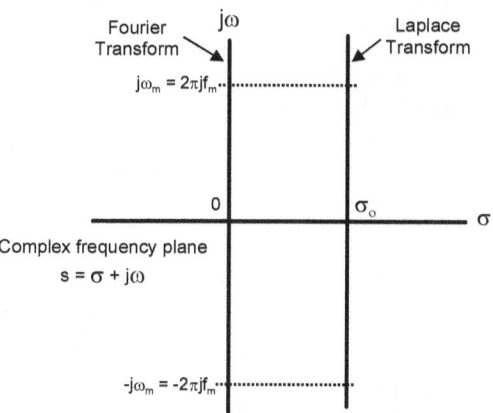

Figure 2. Integration paths along the Bromwich contours for the inverse Fourier and Laplace transforms in the complex s-plane.

The Laplace transform integrals (8) and (9) can be evaluated by direct integration, or as previously noted, by using the FLT which employs the FFT algorithm for performing the necessary integrations. Reference [9] provides a good review of various numerical quadrature methods, as well as an introduction to the FLT approach.

It should be noted in passing that the FLT algorithm for discretely sampled data can be also described by the Chirp z-Transform, as developed by Rabiner in [10]. In this approach, the finite integration path along the Bromwich contour shown in Figure 2 is mapped onto a contour in the complex z-plane. This technique is beyond the scope of this monograph, and the interested reader is referred to Rabiner's work, or to refs. [11 & 12].

2.2. Selection of the Damping Constant σ_o

Not specified in the above discussion is what should be the value of the damping constant σ_o. At a minimum, it should be sufficient to ensure that (1) is satisfied for the given transient waveform being transformed. This implies that each of the "rogue" waveforms of Figure 1 could have a different value of σ_o. Typically the value for the damping constant σ_o is obtained by first examining the early time behavior of the computed inverse Laplace transform of a system response for proper causality. If a null amplitude up to the turn-on time of the response is not noted, then the damping constant should be increased. However, if σ_o is increased too much, there will be exponentially growing noise in the late-time portion of the waveform, and this indicates that the value of σ_o is too large. Thus, there is a trade-off in the accuracy of the early- and late-time portions of the inverse transform.

Two suggested values for σ_o based on the temporal extent of the waveform time window have been made in [6], and these provide starting points for selecting this parameter. These empirical estimates are $\sigma_o = \ln\left(N^2\right)/T_{\max}$ and $\sigma_o = 4\pi/T_{\max}$., where T_{max} is the maximum extent of the transient response time window, and $N = 2^m$, which is the number of waveform sample points.

In our investigations, however, we have found that these values of σ_o provide significant errors at the end of our transient responses. An alternative value that we have found more useful is $\sigma_o = \kappa/T_{\max}$, where the value of κ ranges from 3 to 7, depending on the nature of transient waveforms.

2.3. Determination of a System Response

While (8) and (9) provide the transform relationships between a transient function and its spectrum, it remains to define how the waveform or spectrum would be modified as it passes through a linear system transfer function [13]. Figure 3 shows a block diagram of such a system which has a complex frequency domain transfer function $H(s)$ that relates the output spectrum $G(s)$ to the input spectrum $F(s)$ as

$$G(s) = H(s)F(s) . \qquad (10)$$

In the time domain, this relationship is equivalent to the convolution function between the transient excitation $f(t)$ and the impulse response of the system $h(t)$ as

$$g(t) = \int_0^t h(\tau)f(t-\tau)d\tau$$

$$= \int_0^t f(\tau)h(t-\tau)d\tau \qquad (11)$$

$$\equiv f(t) * h(t)$$

In this expression, the impulse response of the system $h(t)$ is the inverse Laplace transform of $H(s)$, which is given by (9).

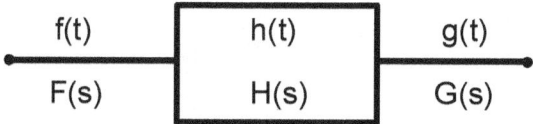

Figure 3. Representation of a linear system with transfer function $H(s)$, excitation spectrum $F(s)$ and response spectrum $G(s)$.

A unit-amplitude transfer function $H(s) = 1$ simply replicates the excitation function $f(t)$ as the response function $g(t)$. The next simplest system transfer function is one that provides a distortionless time delay by the amount t_s between the input and output waveforms, as $g(t) = f(t - t_s)$. In the Fourier transform domain where $\sigma_0 = 0$, this is described by the transfer function $H(\omega) = \exp(-j\omega t_s)$. In the Laplace domain this time-shift transfer function is $H(s) = \exp(-st_s)$, where s is the same complex frequency used in the excitation and response transforms.

For system transfer functions that can be defined analytically in the Fourier domain, extending these into the complex frequency domain is simply done by replacing $j\omega$ by s, as noted in the time-shift example above. Sometimes, however, the Fourier domain transfer function $H(\omega)$ is not represented analytically, but it consists of numerical values that have been obtained as a function of the real (measurable) frequency. This amounts to a Fourier domain transfer function defined along the $j\omega$ axis in the s-plane that must be analytically or numerically continued into the complex frequency plane.

The determination of the Laplace domain transfer function from the Fourier domain transfer function can be done numerically in two different ways. The first method is to recognize that the Laplace spectrum given by (8) is essentially the Laplace transform of the product of two transient functions, $h(t)$ and $\exp(-st)$. Denoting the inverse Fourier transform of $h(t)$ as $H(\omega)$, and noting that the corresponding Fourier spectrum of the exponential function is $1/(j\omega+\sigma_o)$, the convolution theorem [3] permits us to write the Laplace spectrum $H(s)$ as

$$
\begin{aligned}
H(s) &= H(\omega) * \frac{1}{j\omega+\sigma_o} \\
&= \int_{-\infty}^{\infty} H(\xi)\frac{1}{j(\omega-\xi)+\sigma_o}d\xi, \text{ or} \\
&= \int_{-\infty}^{\infty} H(\omega-\xi))\frac{1}{j(\xi)+\sigma_o}d\xi
\end{aligned} \tag{12}
$$

This procedure of (12) is a convolution of two frequency domain spectra, and for spectral responses that have many data points, this calculation canl take a significant amount of time. An alternate and significantly faster approach for determining $H(s)$ from $H(j\omega)$ is to use the inverse Fourier transform operator $\mathcal{F}^{-1}$ of (3) to determine the impulse response $h(t)$, and then to take the direct Laplace transform $\mathcal{L}$ of (8). This procedure is shown symbolically in the following sequence:

$$
H(s) = \mathcal{L}\left[\mathcal{F}^{-1}\left(H(\omega)\right)\right] \tag{13}
$$

Of course, in performing this calculation, the same value of the damping parameter σ_o used for determining the Laplace transform of $f(t)$ should be used in (13).

Chapter 3. Use of the Fast Laplace Transform

In this section we provide several examples of the use of the FLT on waveforms and responses of interest in the EMC field.

3.1.   The FLT Applied to Simple Analytical Waveforms

The difficult waveforms of Figure 1 provide good examples for the use of the FLT. In this case, we assume that the waveforms are passed through a system with a distortionless delay function $H(s) = \exp(-st_s)$ with $t_s = 20$ μs. By applying the FLT with an assumed damping parameter of $\sigma_o = 1.524 \times 10^5$ to the four waveforms, multiplying each by $H(s)$ and then taking the inverse FLT, we obtain the time-shifted waveforms shown in Figure 4. These waveforms are well behaved for both early and late times, and they are clearly a 20 μs shifted replication of the original waveforms. These waveforms provide absolutely no problem for the FLT processing.

To contrast the FLT results with those provided by the FFT operating on the same waveforms in Figure 1, the FLT was re-run with $\sigma_o = 0$, which as we have seen, provides just the FFT. The resulting waveforms are shown in Figure 5. All of these waveforms exhibit the usual fold-through effect at early time, where the non-zero portions of the late-time waveforms appear in early time, when the responses should be identically zero.

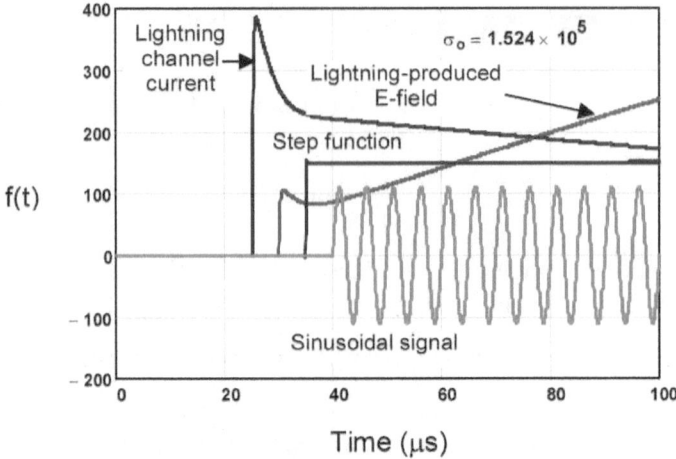

Figure 4. Transient responses of the waveforms of Figure 1 that have been obtained from a 20 μs time-shift operator in the Laplace spectral domain (σ_o = 1.524 x 10^5), using the FLT procedure.

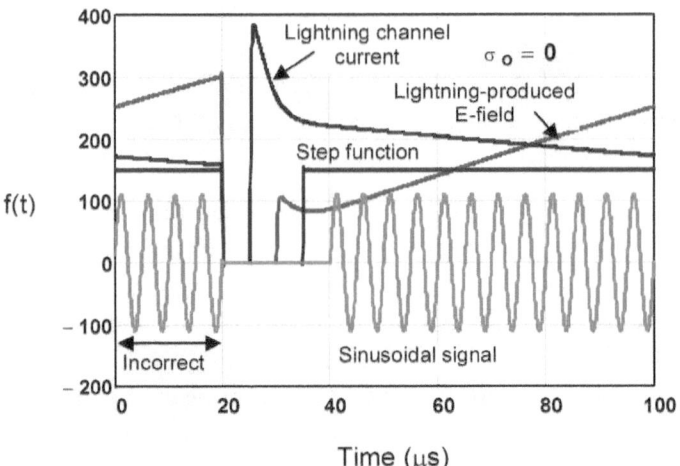

Figure 5. Transient responses of the waveforms of Figure 1 that have been obtained from a 20 μs time-shift operator in the Fourier spectral domain (σ_o = 0), using the FTT.

It is also interesting to examine the behavior of the time-shifted waveforms using the FLT when a larger damping constant is used. Figure 6 shows the waveforms obtained for a damping constant of σ_o = 4.998 x 10^5, which is significantly larger than the suggested values of Section 2.2. We note that the early time behaviors of the shifted waveforms are fine, but at late times there is noise in the response that grows exponentially. This is due to the fact that such a large exponential attenuation of the original time function $f(t)$ in the Laplace transform of (5) has reduced the late time portion of the waveform to essentially the numerical noise level in the computation. Upon taking the inverse transform and multiplying by the growing exponential term of (7), the noise is amplified and it washes out the desired waveform.

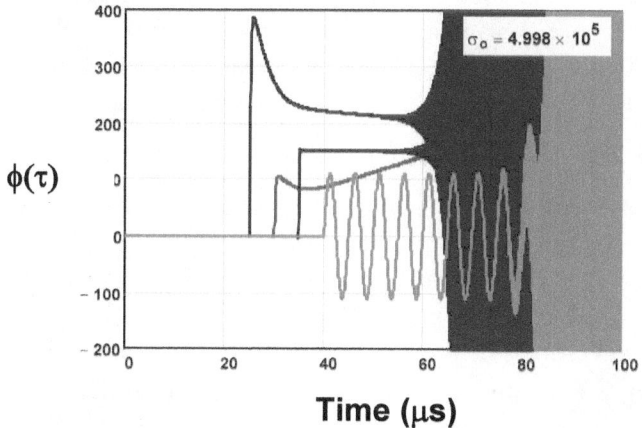

Time (μs)

Figure 6. Time shifted responses of the waveforms of Figure 1 for a large damping constant (σ_o = 4.998 x 10^5) using the FLT.

This divergent behavior of the inverse Laplace transform has been examined in [14], where it is stated that the inversion of the Laplace transform is a good example of an exponentially ill-posed problem. In [14] the authors also use the FFT to compute the direct and inverse Laplace transforms of transient functions, and they discusses the effects that noise and filtering have on the inverse transform. They seem rather pessimistic, however, about the use of the Laplace transform, stating that "Our results give cogent reasons for the general sense of dread most mathematicians feel about inverting the Laplace transform."

Notwithstanding the cautionary remarks in [14], we have noted that the FLT algorithm and its inverse work very well for practical waveforms encountered in the EMC area. Several more examples will be provided in this section.

3.2. Application of the FLT to EMC Problems

The previous section has illustrated the use of the FLT for transient waveforms modified by a very simple system transfer function – a time shift operator. In this section we examine the effects of passing the signal through more complicated transfer functions that arise from EM field coupling to a highly resonant transmission line and EM field interaction with a buried facility that is characterized by a measured continuous wave (CW) transfer function. Finally we conclude with a circuit example that illustrates the use of the FLT on a problem with specified initial conditions.

3.2.1. *A Highly Resonant Transmission Line*

The use of FFT methods for determining the transient responses for the load voltages and currents on the two-wire transmission line of Figure 7 has been described in detail in [15]. The basis of this analysis is the Baum-Liu-Tesche (BLT) equation, which was developed in the time-harmonic (frequency) domain. By substituting s for the variable $j\omega$ in the BLT equation, we can express the load voltages V_1 and V_2 at each end of a lossless transmission line in matrix form as

$$\begin{bmatrix} V_1(s) \\ V_2(s) \end{bmatrix} = \begin{bmatrix} 1+\rho_1 & 0 \\ 0 & 1+\rho_1 \end{bmatrix} \cdot \begin{bmatrix} -\rho_1 & e^{sL/c} \\ e^{sL/c} & -\rho_2 \end{bmatrix}^{-1} \begin{bmatrix} -\dfrac{E^{inc}d}{2}\left(1-e^{sL(1-\cos(\psi))/c}\right) \\ -\dfrac{E^{inc}d}{2}e^{sL/c}\left(1-e^{-sL(1-\cos(\psi))/c}\right) \end{bmatrix}.$$

$$(14)$$

This equation is equivalent to (7.40) in [15], with the simplification that the propagation constant of the transmission line is equal to the free-space propagation constant. In this equation, ρ_1 and ρ_2 are the voltage reflection coefficients at each end of the line, L is the line length, d is the wire separation and E^{inc} is the incident E-field strength. As shown in Figure 7 this incident field arrives at an angle ψ relative to the line orientation. A similar BLT equation can be developed for the load currents, but the details are not presented here.

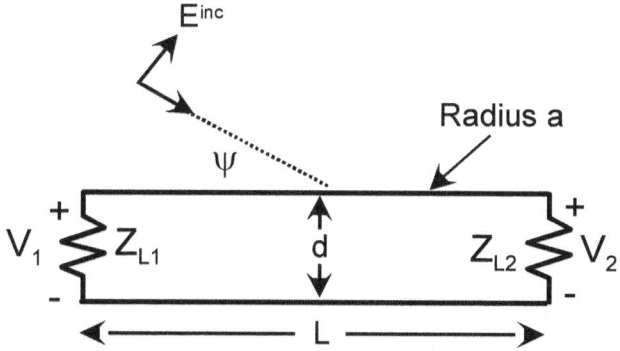

Figure 7. A two-wire transmission line illuminated by an incident plane wave EM field.

As an example of the use of the FLT for this problem, we consider a transmission line having the following dimensions: $L = 1$ m, $d = 1$ cm and conductor radius $a = 1$ mm. The line separation and conductor radius define the characteristic impedance of the line as $Z_c = 120 \ln(d/a) = 276.3\ \Omega$. This impedance is used with the load impedances to determine the voltage reflection coefficients, as $\rho = (Z - Z_c)/(Z + Z_c)$.

The excitation of this transmission line is provided by the 100 V/m exponential waveform shown in Figure 8. This waveform has a 1 μs decay time and a waveform start time of 0.1 μs. For this example, the angle of incidence of the excitation field is $\psi = 45°$.

In this example, we calculate the voltage at load #2 and the current at load #1 for a highly resonant line. This is difficult to do using conventional FFT methods. To obtain such a resonant line, we will use a short circuit for the left load ($Z_1 = 0$) and an open circuit for the right load ($Z_2 = \infty$).

15

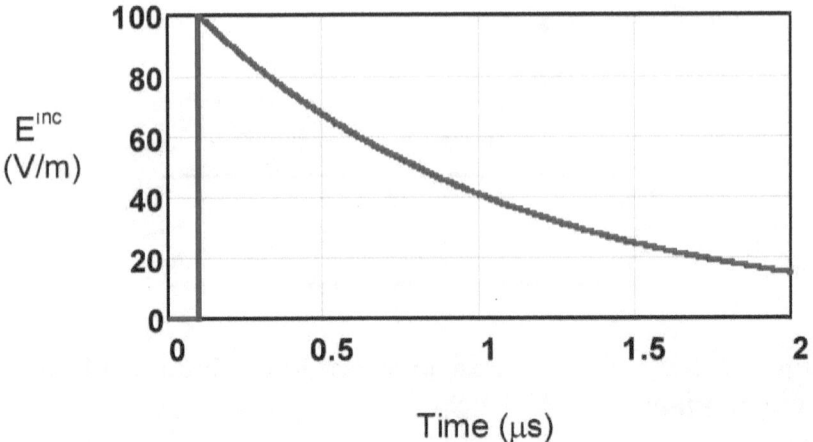

Figure 8. The incident E-field excitation waveform for the transmission line.

Reference [15] discusses how transient responses of highly resonant structures like this example can be solved in the spectral domain by expanding the inverse matrix in (14) in a suitable manner and then computing a modified spectrum using only a few terms of the expansion. The FFT of this modified spectrum provides a waveform that is correct for several oscillations, but then goes to zero for later times. This property of the waveform permits the use of the FFT in the analysis.

If this spectral modification procedure is not used and the voltage or current spectra are calculated along the $j\omega$ axis of Figure 2, there are periodic singularities that occur in the spectra, due to the resonances on the transmission line. These are shown in the spectral plot of Figure 9. If the FFT is applied to these spectra, incorrect waveforms like that shown in Figure 10 result. Note that the amplitude of this waveform is unreasonably large and the waveform is non-causal, as there is a response prior to the turn-on time of the excitation at 0.1 μs.

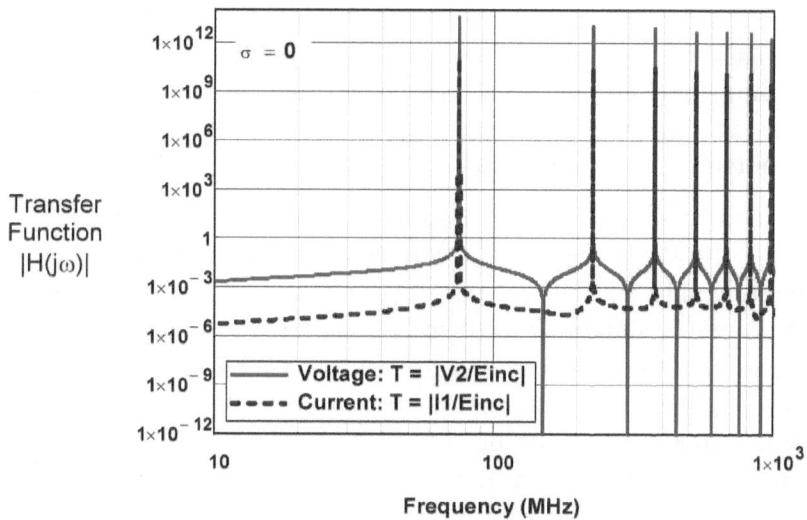

Figure 9. Plots of the voltage and current transfer function magnitudes $|V_2/E^{inc}|$ and $|I_1/E^{inc}|$ along the $j\omega$ axis ($\sigma_o = 0$).

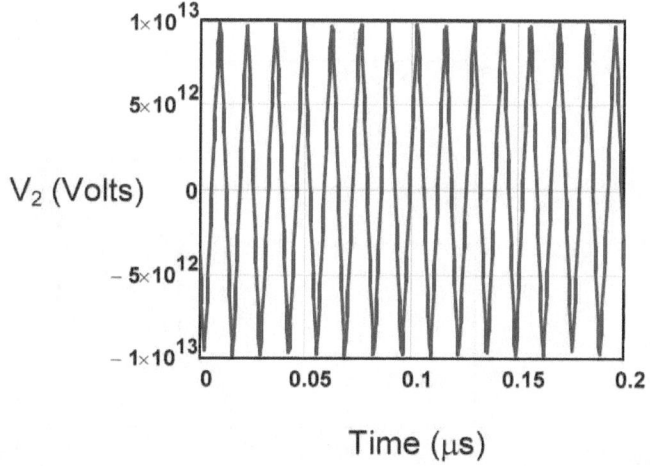

Figure 10. The transient response for V_2, as computed from the voltage transfer function of Figure 9 using the FFT. (Incorrect)

The waveform is physically incorrect for two reasons. The first is because we have neglected possible loss in the transmission line, the line resonances are infinite. In reality, loss will provide a finite Q to these singularities, and the response waveform in Figure 10 will eventually die out. The second reason is due to the fact that we are sampling the FFT spectrum only at discrete points. If insufficient points are taken, the spectral peaks are not well-sampled and the computed transient response is erroneous.

Thus, to calculate this response accurately using an FFT, loss must be added to the problem, and more points added to the waveform and its spectral response. Changing the Z_l load to 1 Ω will provide a suitable loss to the structure, and increasing the time window of the exponential excitation waveform to 6 μs will increase it sufficiently so that the response of the line has died out. Then, increasing the number of points in the waveform from 4086 to 32,768 will provide for better sampling of the resonances.

Figure 11 shows the resulting early-time response for the voltage V_2 with these modifications to the problem parameters, for the FFT processing. Aside from the very slight decay in the oscillations on the line due to the added loss, the response is a reasonable approximation for this highly resonant line.

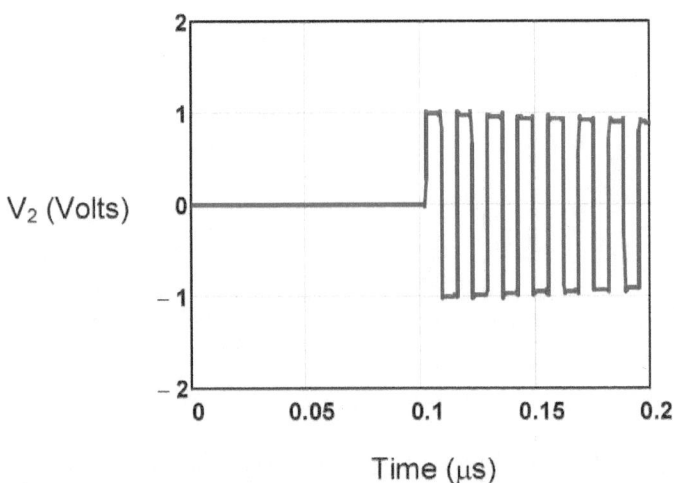

Figure 11. Plot of the transient voltage response V_2 using the FFT by adding a small amount of loss, extending the waveform window and increasing the number of points in the waveform and spectrum.

Alternatively, the FLT can be applied to the computed Laplace spectrum for a suitable value of $\sigma_o = 3.612$ x 10^6 (and with the smaller sampling density of only 4096 points in the transient waveform). Figure 12 plots the Laplace spectral magnitudes for the voltage and current transfer functions along the Bromwich contour $\sigma_o + j\omega$. Note that in Figure 12 the abscissa variable is the frequency $f = \omega/2\pi$, which is derived from the imaginary part of the complex frequency s. Figure 13 presents the computed transient voltage V_2 for both early and late times.

The short circuit current I_1 has been calculated in a similar manner using the BLT current equation, and its transient response using the FLT is shown in Figure 14.

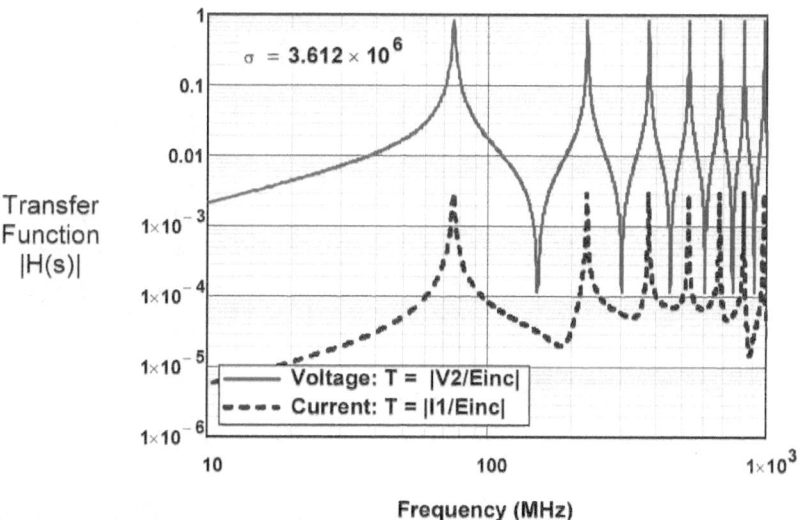

Figure 12. Plots of the voltage and current transfer function magnitudes $|V_2/E^{inc}|$ and $|I_1/E^{inc}|$ along the Bromwich contour $\sigma_o + j\omega$.

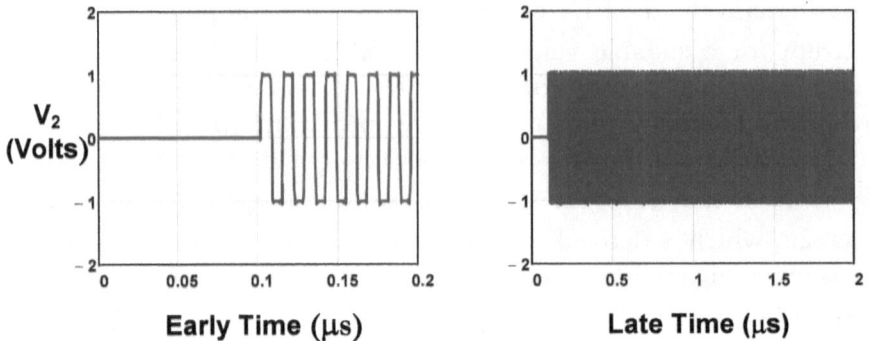

Figure 13. Plots of the early time (left) and late time (right) waveforms for the voltage V_2 on the lossless transmission line using the FLT.

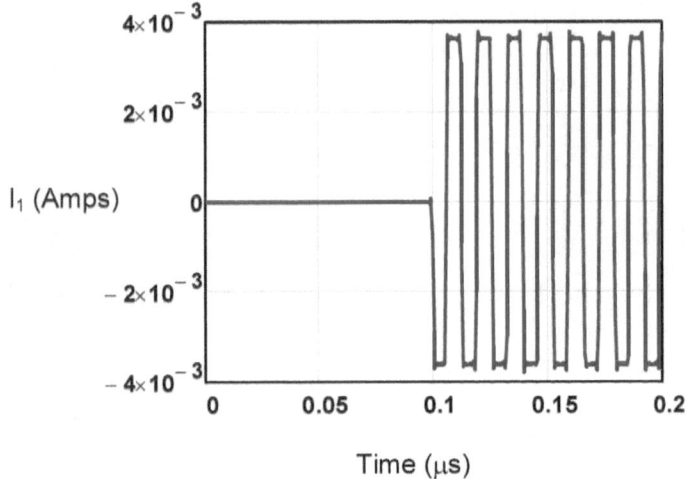

Figure 14. Plots of the early time current waveform I_1 using the FLT.

It should be pointed out that this FLT analysis approach also can be applied to transmission lines having loss, with no undue difficulties. This is unlike the more conventional analysis of Chang and Kang [16].

3.2.2. *Use of the FLT with a Measured CW Transfer Function*

Measured CW (time harmonic) transfer functions are often used to characterize a linear time invariant system. Such transfer functions have both a magnitude and phase, and if these functions are known over a sufficiently large frequency range, the transient response of the system to an arbitrary transient excitation can be computed.

One example of such a transfer function has been described in [17], where radiated field measurements were made on an underground bunker. The facility was illuminated by a wide-band antenna, and measurements of the transfer function between a component of the internal magnetic field and the excitation voltage of the antenna were made. A network analyzer was used to measure both the magnitude and phase of the transfer function.

Figure 15 shows the magnitude of a typical transfer function measured from 1 MHz to about 1 GHz. The corresponding phase for this transfer function was also measured and is used in the calculations, but it is not displayed. As noted earlier, this transfer function is measured along the $j\omega$ axis of Figure 2 and for use with the FLT. It must be analytically continued onto the Bromwich contour $\sigma_o + j\omega$. This will be done using the procedure shown in (13).

Figure 16 shows the computed impulse response $h(t)$, which is computed by an inverse FFT applied to the spectrum of the measured transfer function of Figure 15. For accurate processing, such transients should always be checked for causality and for proper behavior at late times. This particular transient is well behaved – indicating that the measured CW data are sufficiently accurate.

For the second step in evaluating (13), the FLT is applied to $h(t)$ with an assumed damping constant $\sigma_o = 6.62 \times 10^6$. This provides the value of the transfer function along the Bromwich contour $\sigma_o + j\omega$. Figure 17 provides an overlay plot of the measured and analytically extended FLT spectra.

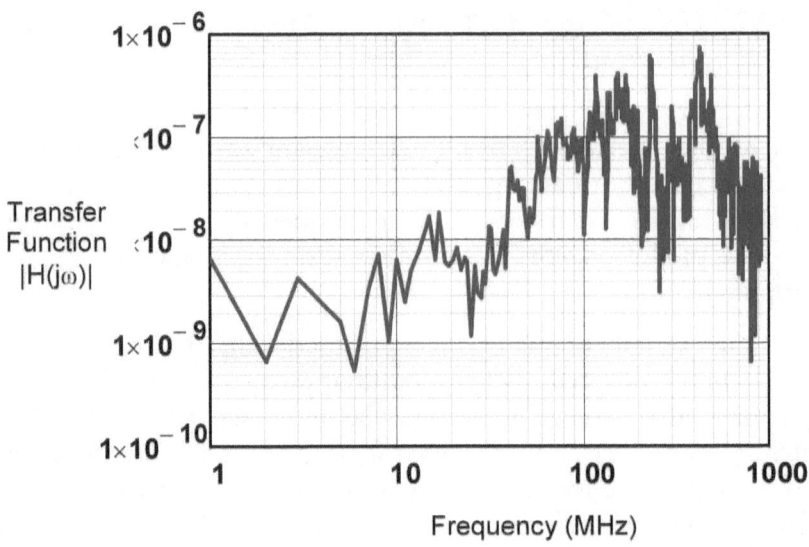

Figure 15. A typical measured transfer function magnitude for the internal magnetic field in a buried facility [17].

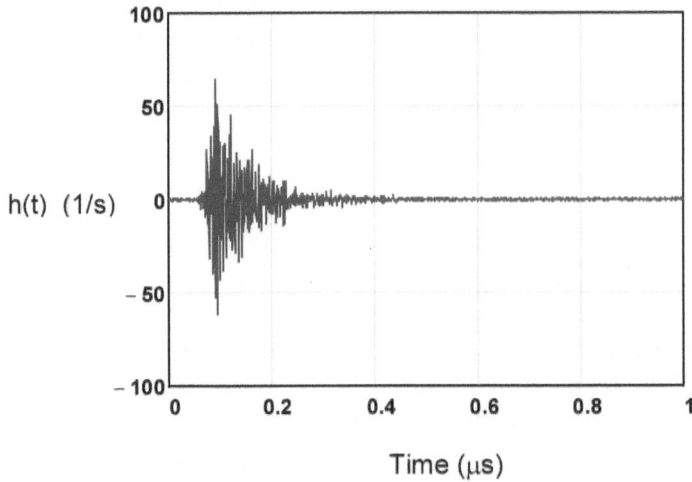

Figure 16. Inverse FFT of the measured complex CW transfer function $H(\omega)$ of Figure 15, yielding the system impulse response $h(t)$.

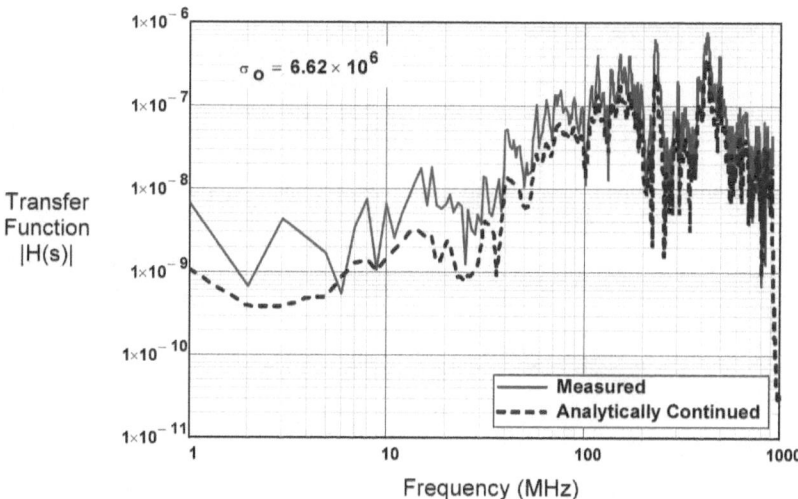

Figure 17. Overlay plot of the measured transfer function magnitude with the analytically continued transfer function on the Bromwich contour $\sigma_o +j\omega$.

To illustrate the FLT processing of this transfer function, we assume that the antenna is excited by a unit-amplitude time-shifted sinusoidal waveform

$$f(t) = e^{-\alpha(t-t_s)} \sin(2\pi f_o(t-t_s)\Phi(t-t_s) \text{ volts}$$

with $f_o = 100$ MHz, $\alpha = 1 \times 10^{-7}$ (s^{-1}) and $t_s = 0.05$ µs. Figure 18 plots this waveform and its computed FFT and FLT spectra.

To determine the internal transient response for the magnetic field for this particular antenna excitation, we could convolve the waveforms in Figure 16 and Figure 18, but this would take a significant amount of computation time. Alternatively, we can multiply either the FLT or the FFT spectra of the excitation by the appropriate transfer functions of Figure 17 and take the inverse FLT or FFT to obtain the transient responses.

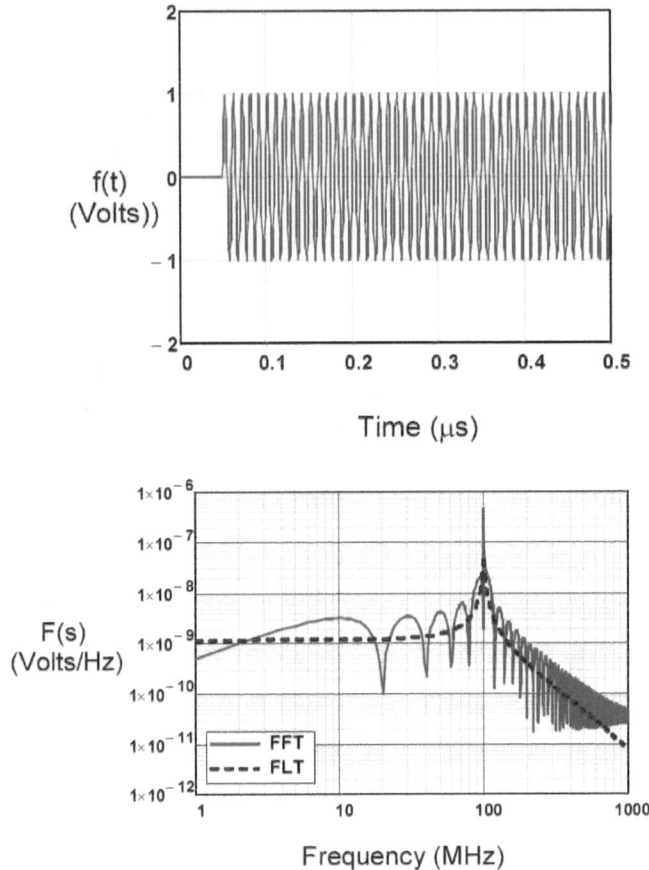

Figure 18. The assumed transient voltage source of the radiating antenna (top) and the corresponding FLT and FFT spectral magnitudes (bottom). (σ_o = 6.62x10^6 for the FLT spectrum).

Figure 19 presents the transient response resulting from the FLT processing using σ_o = 6.62 x 10^6. The late time response does not show the exponential growth of the noise, so the damping parameter is appropriate. There is, however, some very slight non-causal behavior in the early time response, which is attributed to the fact that the measured spectrum had to be interpolated to correspond to the frequencies of the FLT spectrum of the excitation. This interpolation of the measured data introduces small errors in the measurements that give rise to this anomalous, but small effect.

In examining the turn-on time of the waveform in Figure 19, we note that the antenna excitation waveform turns on at 0.05 μs, and the response in the facility has an additional time delay of about 0.07 μs. This provides a total time delay of 0.12 μs, which is indicated in the figure.

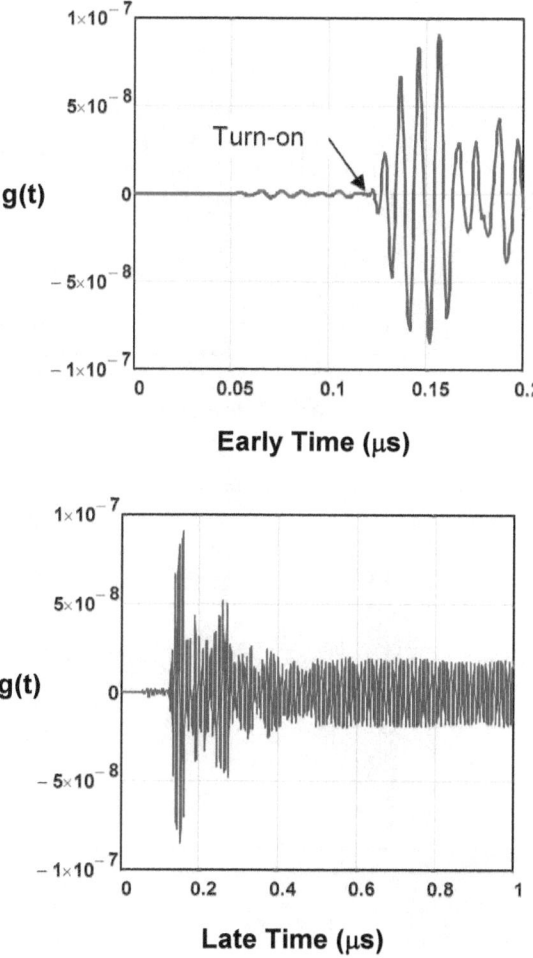

Figure 19. Early- and late-time calculated transient responses of the internal magnetic field for the antenna excitation of Figure 18 using the FLT processing.

Figure 20 shows the resulting response waveform for the FFT processing of the spectra. It is clear that the response waveform is not causal. Furthermore the time from the transition of the very early non-causal sinusoidal waveform to the apparent system response waveform is not correct value of 0.12 μs. This FFT calculated response is clearly incorrect, and it shows the difficulties encountered in trying to use an FFT process in this case.

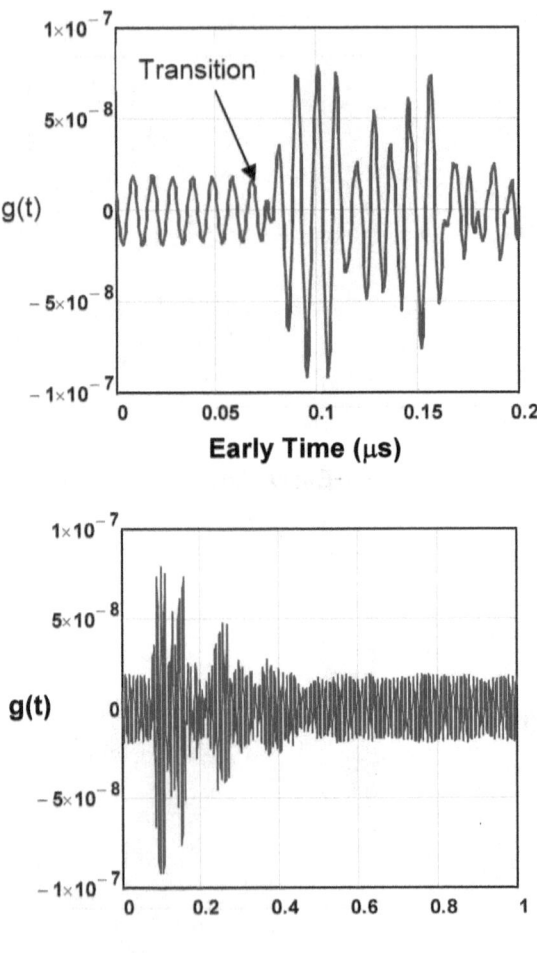

Figure 20. Early- and late-time calculated transient responses of the internal magnetic field for the antenna excitation of Figure 18 using the FFT.

3.2.3. *Use of the FLT on Problems with Initial Conditions*

A very useful attribute of the Laplace transform is its ability to include initial conditions in the analysis. This benefit is also found if the transforms are performed numerically using the FLT. To illustrate this, we consider a Marx pulser model with a peaking circuit, as shown in Figure 21. The use of this circuit for producing fast-rising pulses has been discussed by Taylor and Giri [18] and we will illustrate the use of the FLT for conducting an analysis of this generator.

The circuit consists of a Marx pulser, having circuit elements R, L and C, and a peaker capacitance C_p as a load. The secondary peaking circuit contains a load inductance L_p (which is usually a parasitic element) and a load impedance Z_L. Initially, the capacitance C is charged to a voltage $v_c(0+) = -V_o$, with the other capacitor and inductors having zero initial voltage and current. At $t = t_1$, the Marx switch is closed and the loop current $i_1(t)$ begins to flow in the Marx section[4]. The current $i_1(t)$ charges the peaker capacitance and when the charge is sufficient, the peaker switch is closed at time $t = t_2$. This results in a very fast rising current in the peaker circuit and a transient excitation of Z_L.

For this example we assume the following parameters: $L = 3$ μH, $L_p = 1$ nH, $C = 5$ nF, $C_p = 1$ nF and $R = 0.1$ Ω.

Two different values for the peaker load Z_L are considered: 1) a fixed load of 50 Ω, and 2) a load provided by the input impedance of a lossless 50 Ω transmission line with a length of 60 meters and loaded by a 20 Ω resistance at the end.

The switching times for this circuit are assumed to be $t_1 = 0.1$ μs, and $t_2 = 0.2$ μs. These parameters are not optimized for a particular pulser design, but are used only for illustrative purposes.

---

[4] In this development, we will use lower case variables like $v(t)$ and $i(t)$ to denote transient responses, and upper case variables $V(s)$ and $I(s)$ to denote the Laplace transform responses.

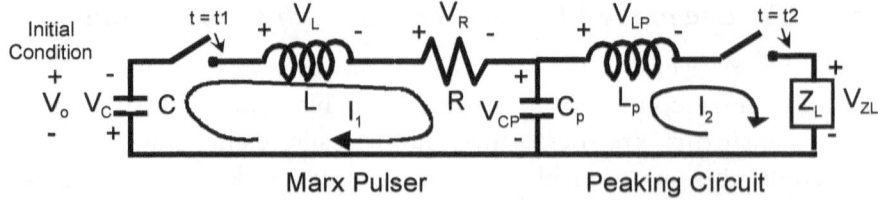

Figure 21. A Marx generator and peaking circuit used to illustrate the use of initial conditions in the FLT.

The Laplace transform analysis circuits having initial conditions is standard material in circuit analysis texts. In [13] for example, it is shown that if $\mathcal{L}(f(t)) = F(s)$ represents the Laplace transform operation of (8), the transform of the derivative of $f(t)$ is given by

$$\mathcal{L}\left(\frac{df}{dt}\right) = sF(s) - f(0_+) , \tag{15}$$

where $f(0_+)$ is the initial value of the transient response as $t \to 0$, as approached from $t > 0$.

Applying (15) to an inductor and capacitor, which are described by the relationships $v(t) = L \, di / dt$ and $i(t) = C \, dv / dt$, respectively, the V-I relationships for these elements in the Laplace domain are

$$V(s) = sLI(s) - Li(0_+) \text{ (for the inductor) and} \tag{16a}$$

$$I(s) = sCV(s) - Cv(0_+) \text{ (for the capacitor).} \tag{16b}$$

After the Marx switch is closed, and prior to the peaker switch being closed at $t = t_2$, Kirchhoff's voltage law (KVL) may be applied to the current loop in the Marx section to determine the current I_1. As this requires the sum of all of the voltages across the circuit elements in the loop be zero, it is convenient to invert (16b) to obtain the capacitor voltage in terms of the current, which is

$$V(s) = \frac{I(s)}{sC} + \frac{v(0_+)}{s} \text{ (for the capacitor).} \tag{17}$$

Thus, the KVL for the Marx section becomes

28

$$\left(\frac{I_1(s)}{sC} + \frac{v_C(0_+)}{s}\right) + \left(sLI_1(s) - Li_1(0_+)\right) + RI_1(s) + \left(\frac{I_1(s)}{sC_p} + \frac{v_{Cp}(0_+)}{s}\right) = 0$$

(18)

Since the initial current $i_1(0_+)$ and peaker capacitance voltage $V_{Cp}(0_+)$ are both zero, and since the initial Marx capacitor voltage is $v_c(0+) = -V_o$, the loop current in (18) can be expressed as

$$I_1(s) = \frac{\dfrac{V_o}{s}}{sL + \dfrac{1}{s}\left(\dfrac{1}{C} + \dfrac{1}{C_p}\right) + R}.$$

(19)

Equation (19) represents spectrum for the waveform that starts at $t = 0$. The current spectrum can be modified by the time shift Laplace operator $\exp(-st_1)$ to account for the Marx switching time at $t = t_1$.

The usual way of taking the inverse transform of (19) to obtain the transient current in the Marx circuit is to expand the equation into a partial fraction expansion and then take an analytical inverse transform of each term. Alternatively, we can obtain a numerical solution by using the FLT on the Laplace spectrum of the current.

Without the peaker switching, the transient current $i_1(t)$ will oscillate for a long time until it is eventually damped out by the resistance R. When the peaker switch is activated, however, a second current loop is introduced into the circuit and this must be included in the circuit analysis. This is done by developing two coupled mesh equations for the spectral currents $I_1(s)$ and $I_2(s)$.

These equations involve the current through L and the voltages across C and C_p at time t_2, as initial conditions.

The two mesh equations can be expressed in matrix form as

$$\begin{bmatrix} sL + \dfrac{1}{s}\left(\dfrac{1}{C} + \dfrac{1}{C_p}\right) + R & -\dfrac{1}{sC_p} \\[2ex] -\dfrac{1}{sC_p} & sL_p + \dfrac{1}{sC_p} + Z_L \end{bmatrix} \cdot \begin{bmatrix} I_1 \\ I_2 \end{bmatrix} = \begin{bmatrix} Li_1(t_1) - \dfrac{1}{s}\left(v_C(t_1) + v_{Cp}(t_1)\right) \\[2ex] \dfrac{v_{Cp}(t_1)}{s} \end{bmatrix}$$

(20)

where the right hand vector is the forcing function, and the 2x2 matrix to the left is the system impedance matrix. This equation can be inverted either numerically or analytically to permit the evaluation of the currents I_1 and I_2.

The resulting transient currents flowing in the pulser circuit were determined by first defining a maximum time window desired for the responses, which was chosen to be 20 μs. With a choice of $2^{13} = 8192$ sample points, a sampling interval of 2.44 ns is obtained. The value for the damping parameter of the Laplace transform in (5) was chosen as $\sigma_o = \kappa / T_{max}$, with $\kappa = 4$. This gives $\sigma_o = 0.2 \times 10^6$.

The transient response for $i_1(t)$ was computed from the inverse FLT of (19). Figure 22 shows this current starting at 0.1 μs and continuing as a slowly damped sine. The inverse FLT of $I_1(s)$ from (20) was computed and shifted to begin at 0.2 μs, just after the peaking circuit switches. This response is the dotted line in Figure 22, and it shows how the Marx current changes when the switching occurs. The other solid curve in this figure represents the current flowing in the peaker circuit, $i_2(t)$. This response is seen to have a very fast rise time, which is controlled by the load resistance and the peaker inductance L_p. For this calculation, the peaker circuit load was taken to be the 50 Ω resistive element.

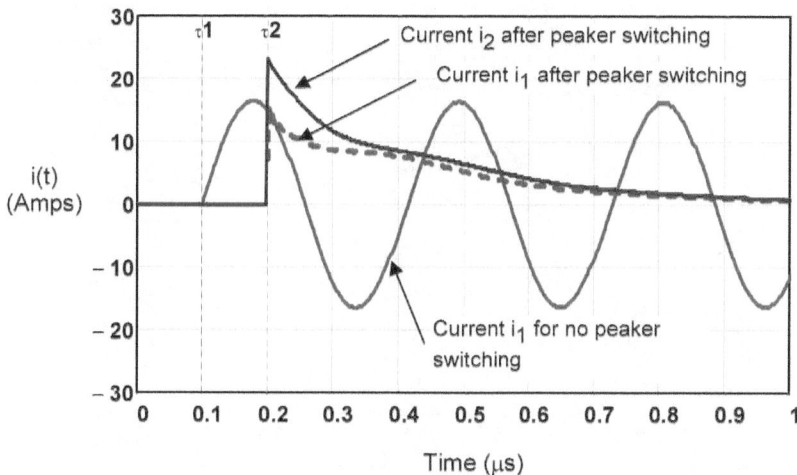

Figure 22. Transient response of the Marx generator current $i_1(t)$ assuming no peaker switching, and the currents $i_1(t)$ and $i_2(t)$ for peaker switching at 0.2 μs. (Peaker load is 50 Ω.)

Once the current spectral responses are determined, the voltages across the capacitors can be determined using (17). Note that this requires including the initial conditions of the capacitor voltages in the inverse FLT. Figure 23 illustrates the transient voltages V_C and V_{Cp}, as computed using the FLT. The solid curves represent the responses if no switching of the peaker circuit occurs, while the dotted curves show the behavior of the voltages after the switching at $t = t_2$.

While the previous results could have been obtained analytically using the conventional partial fraction expansion method, it is more difficult to do this if the load impedance Z_L is a measured function, as in the previous example, or if it is a non-rational function of frequency. In these cases, the use of the FLT makes the analysis significantly easier.

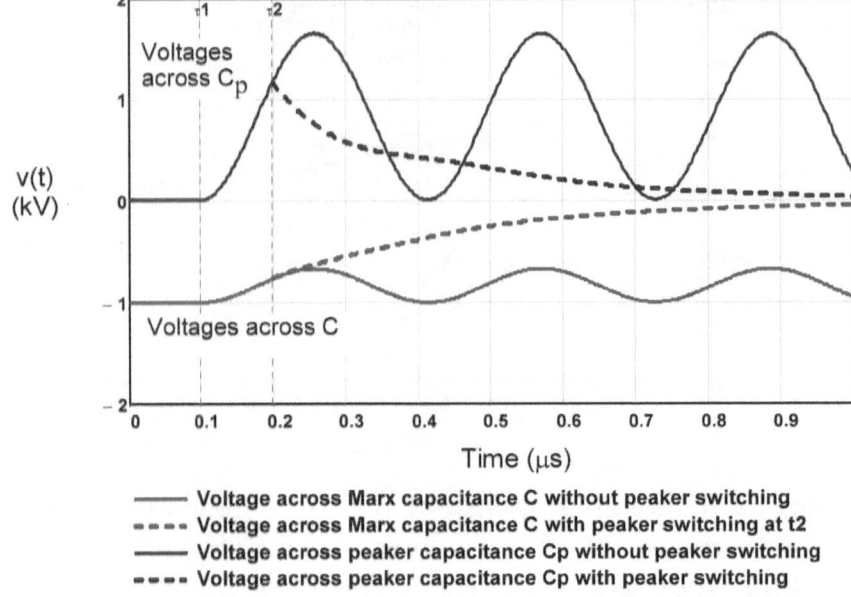

Voltages
across C_p

$t1$ $t2$

v(t)
(kV)

Voltages across C

Time (μs)

───── Voltage across Marx capacitance C without peaker switching
▬ ▬ ▬ Voltage across Marx capacitance C with peaker switching at t2
───── Voltage across peaker capacitance Cp without peaker switching
▬ ▬ ▬ Voltage across peaker capacitance Cp with peaker switching

Figure 23. Plots of the transient voltages across C and C_p, with and without the switching of the peaker circuit at $t = t_2$. (Peaker load is 50 Ω.)

As an example, we consider the case of a section of a 60 meter, 50 Ω transmission line connected to the pulser instead of the 50 Ω resistance. The input impedance of this line, when terminated in the 20 Ω load, can be calculated from the transmission line models given in [15]. This input impedance is shown by the solid curve in Figure 24. The analytical continuation of this impedance function onto the $\sigma_o = 0.2 \times 10^6$ contour in the complex s-plane can be done as in the previous example through the use of (13), and this is also shown in the figure.

Figure 25 presents the impulse response of the input impedance of the transmission line. This clearly exhibits the multiple reflections occulting on the transmission line.

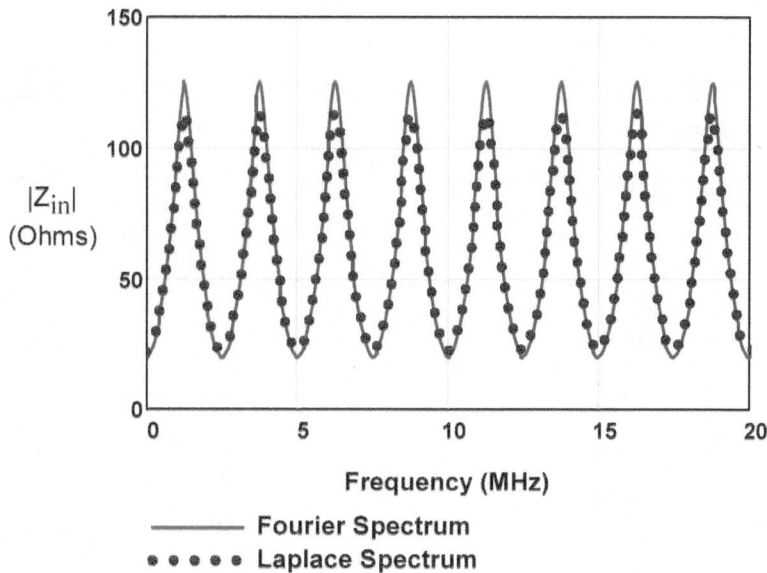

Figure 24. Plot of the input impedance magnitude $|Z_{in}(j\omega)|$ and the Laplace spectral magnitude $|Z_{in}(s)|$ for the 60 meter transmission line load.

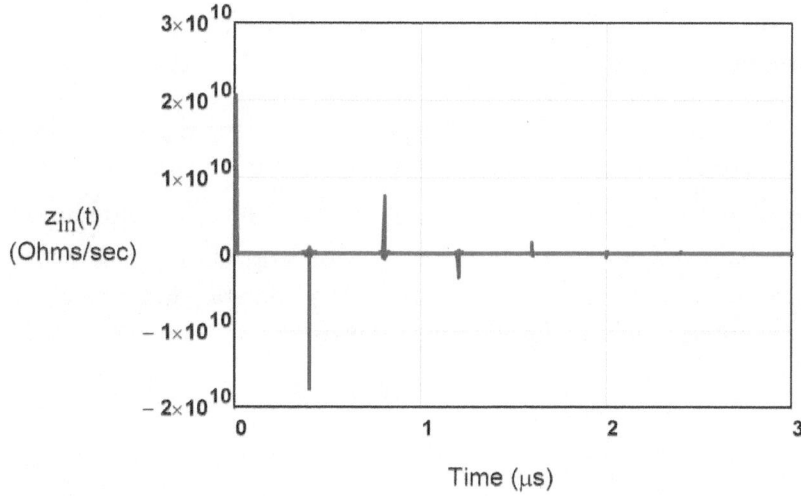

Figure 25. The impulse response of the impedance function, as determined by an inverse FFT of the computed function $Z_{in}(\omega)$.

Figure 26 plots the Laplace spectra for the voltage across the load element Z_L as a function of frequency for both the transmission line load and the constant 50 Ω load. The presence of the oscillations on the transmission line is evident.

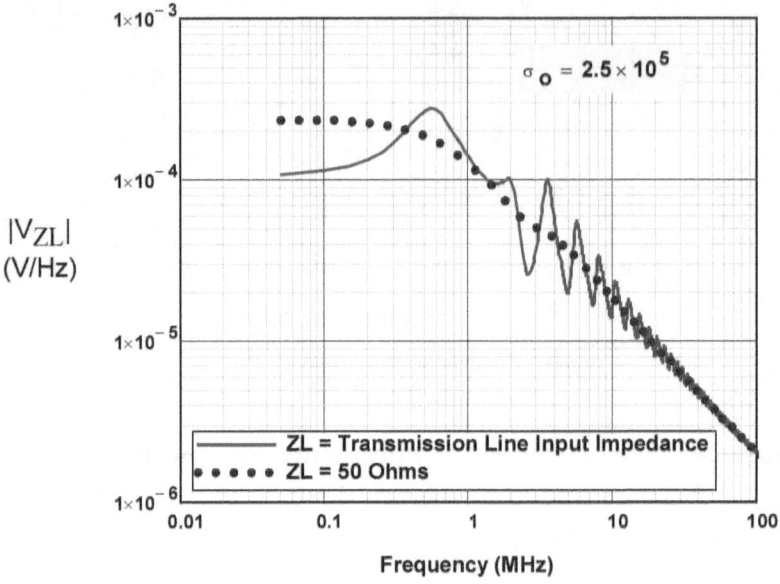

Figure 26. Plot of the load voltage Laplace spectra for the transmission line load and the 50 Ω load.

Figure 27 shows the corresponding transient responses for the load currents, which have been time shifted to start at time t_2. It is seen that for a time $t = 2 \times 60/c = 0.4$ μs after the switching at t_2 (or at 0.6 ms absolute time) the waveform appears like that across a constant 50 Ω load. However, after the end reflection arrives back to the pulser circuit, the response deviates from the constant load value.

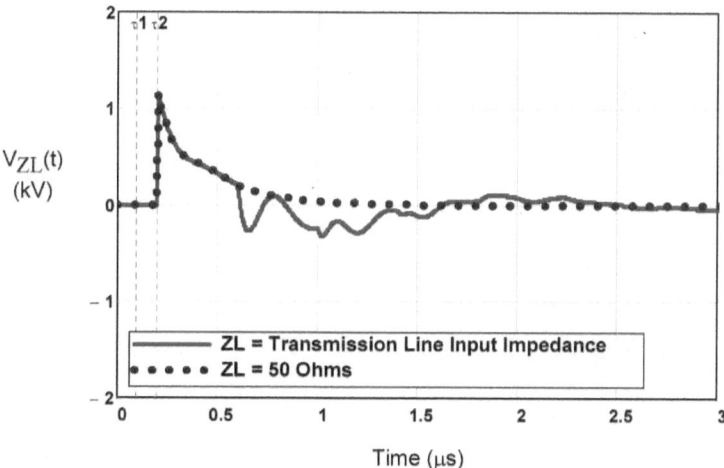

Figure 27. Plot of the transient load voltage across Z_L for the transmission line load and the 50 Ω load.

Chapter 4. Summary

This monograph has reviewed the use of the fast Laplace transform, as implemented using the fast Fourier transform. This method has been described earlier by investigators in the electrical power community, but it does not seem to be widely used in the EMC community.

We have illustrated the use of the FLT for time shifting several waveforms that pose difficulties when used with the FFT. This time shift procedure is performed in the spectral domain by multiplying the Laplace spectrum by the time shift spectral operator, and then inverting back into the time domain with the inverse fast Laplace transform. The use of this method on a highly resonant transmission line and on measured CW transfer function data also has been discussed.

Mathematicians have described the inverse Laplace transform as being "ill conditioned" and refer to this as a "bad truth" about the procedure. In our work, we have observed that this ill conditioned nature of the Laplace inversion occurs in the form of exponential noise in the late time portion of the inverted transient response. The reason for this ill conditioning is similar to trying to find the limit of $\sin(x)/x$ numerically. Eventually this limit becomes noise dominated and the actual limit of unity is never found.

In the inverse Laplace transform case, almost all computed time domain functions will become unstable if the time is allowed to run long enough. Hence, this method is "bad" because it does not replicate the actual waveform in late times. However, this procedure does not limit the utility of the method for early and intermediate times, as noted in the examples presented here.

The key to obtaining useful Laplace transform results is determining a suitable value for the damping constant, σ_o. As noted here, this parameter must be positive and its "optimal" value cannot be defined explicitly. Several suggested values for σ_o have been given for waveforms based on their time duration or their sampling rate. But ultimately, it is up to the user of this procedure to decide upon a suitable value.

Different values of σ_o will provide different responses of the inverse transform at late times, and perhaps this ambiguity in the inversion process is what upsets the mathematicians.

Chapter 5. References

1. Heideman, M. T., D. H. Johnson, and C. S. Burrus, "Gauss and the History of the Fast Fourier Transform," *IEEE ASSP Magazine*, 1 (4), 14–21 (1984).

2. Cooley, James W., and John W. Tukey, "An Algorithm for the Machine Calculation of Complex Fourier Series", *Mathematics of Computation*, Vol. 19, pp 297–301 (1965).

3. Papoulis, **The Fourier Integral and its Applications**, McGraw-Hill, New York, 1962.

4. Nucci, C. A., et. al., "Lightning-Induced Voltages on Overhead Lines", *IEEE Trans. EMC*, Vol. 35, No.1, February 1993.

5. Ramirez, A., P. Gomez, P. Moreno, and A. Gutierrez, "Frequency Domain Analysis of Electromagnetic Transients through the Numerical Laplace Transform", *Proceedings of the 2004 IEEE Power Engineering Society General Meeting*, 10 June 2004, pp 1136 – 1139.

6. Gómez Zamorano, P. and F. A. Uribe Campos, "On the Application of the Numerical Laplace Transform for Accurate Electromagnetic Transient Analysis", *Revista Mexicana de Física*, 52(3), Junio 2007, pp. 198-204.

7. Day, S. J., J. Battisson, N. Mullineux, and J. R. Reed, "Developments in Obtaining Transient Response using Fourier Transforms, Part III: Global Response", *Int. J. Elect. Engng. Educ.*, Vol. 6, pp 259-265, 1968.

8. LePage, W. R., Complex Variables and the Laplace Transform for Engineers, Dover Publications, New York, 1961.

9. Cohen, A. M., **Numerical Methods for Laplace Transform Inversion**, Springer Science + Business Media, New York, 2007.

10. Rabiner, L. R., F. W. Schafer, and C. M. Rader, "The Chirp z-Transform Algorithm", *IEEE Trans. Audio Electroacoust.*, Vol. AU-17, June 1969.

11. Oppenheim, A. V., and R. W. Schafer, **Digital Signal Processing**, Sec. 6.6, pg. 321, Prentice Hall, Inc. New Jersey, 1975.

12. Rabiner, L., "The Chirp z-Transform Algorithm—A Lesson in Serendipity", *IEEE Signal Processing Magazine*, 1053-5888/04/, pp 118-119, March 2004.

13. Cheng, D. C., **Analysis of Linear Systems**, Addison-Wesley Publishing Co., Reading, Mass., 1959.

14. Epstein, C. L. and J. Schotland, "The Bad Truth about Laplace's Transform", *SIAM Review*, Issue 3, pp 504-520, August 2008. (Document is available at the internet address http://www.math.upenn.edu/~cle/papers/laplce_rev2.pdf).

15. Tesche, F. M., et. al, **EMC Analysis Methods and Computational Models**, John Wiley and Sons, New York, 1997.

16. Chang, E. C. and S-M Kang, "Computationally Efficient Simulation of a Lossy Transmission Line with Skin Effect by Using Numerical Inversion of Laplace Transform", *IEEE Trans Circuits and Systems*, Vol. 39, No 11, Nov. 1992.

17. Tesche, F. M., et. al., "Measurements of High-Power Electromagnetic Field Interaction with a Buried Facility", *Proceedings of the International Conference on Electromagnetics in Advanced Applications*, Torino, Italy, Sept.10-14, 2001.

18. Taylor, C. D., and D. V. Giri, **High Power Microwave Systems and Effects**, Taylor and Francis, Washington DC, 1994.

Chapter 6. Index

www.ingramcontent.com/pod-product-compliance
Lightning Source LLC
Chambersburg PA
CBHW021042180526
45163CB00005B/2238